Joseph Rufus Hunter

The relation of the anilides of orthosulphobenzoic acid

Joseph Rufus Hunter

The relation of the anilides of orthosulphobenzoic acid

ISBN/EAN: 9783337305970

Printed in Europe, USA, Canada, Australia, Japan

Cover: Foto ©berggeist007 / pixelio.de

More available books at **www.hansebooks.com**

.:- THE RELATION OF THE ANILIDES -:-

OF

ORTHOSULPHOBENZOIC ACID.

Dissertation

Submitted to the Board of University Studies of the

Johns Hopkins University for the Degree

of Doctor of Philosophy

by

JOSEPH RUFUS HUNTER.

1895.

CONTENTS.

ACKNOWLEDGMENT.

The investigation described in the following pages was undertaken at the suggestion of Professor Ira Remsen, and carried on under his constant supervision. I gladly avail myself of this opportunity to express to him, as well as to Professor H. N. Morse, my sincere gratitude for the assistance they have freely rendered me during its progress.

I. INTRODUCTION.

Orthosulphobenzoic acid was first prepared by Remsen and Fahlberg (1) in the year 1879, by oxidizing orthotoluene-sulphoneamide by potassium permanganate in alkaline solution. It was further investigated by Remsen and Dohme (2), who prepared a large number of its salts, as well as the chloride obtained by the action of phosphorus pentachloride on the acid ammonium salt.- They obtained crystals of the chloride; and from its ability to form chloro-ethers of the general formula,

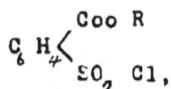

$$C_6 H_4 \Big\langle \begin{matrix} Coo\ R \\ SO_2\ Cl, \end{matrix}$$

in which the chlorine could be easily replaced by hydroxyl forming an acid of the structure, $C_6 H_4 \Big\langle \begin{matrix} Coo\ R \\ SO_2\ O\ H, \end{matrix}$ they represented its structure by the formula,

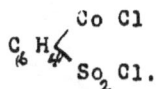

$$C_6 H_4 \Big\langle \begin{matrix} Co\ Cl \\ So_2\ Cl. \end{matrix}$$

No evidence was obtained that this chloride was not homogeneous.-

Remsen and Coates (3), in 1891, studied this action of aniline on the chloride, and obtained two isomeric anilides quite different in properties, and an anil. From this work it seemed

(1) American Chemical Journal 1, 433.
(2) Ibid, 11, 332.
(3) Ibid, 17, 311.

quite evident that there were two isomeric chlorides, from
which the anilides were derived.

Later Remsen and Kohler (1) obtained one of these
chlorides pure, in crystalline form, while the other one was an
oil, always containing more or less of the crystalline variety.
It was further shown in this investigation that the anil and
the fusible anilide described by Remsen and Coates (2), were
derivatives of the symmetrical (or solid) chloride, $C_6H_4\begin{smallmatrix}CoCl\\SO_2Cl\end{smallmatrix}$,
while from the unsymmetrical or liquid chloride, for which the
formula $C_6H_4\begin{smallmatrix}CCl_2\\SO_2\end{smallmatrix}>O$ had been proposed, the infusible anilide,
together with some of the fusible one, was obtained.- The
liquid chloride was first obtained pure by Remsen and Saunders
(3), who obtained it as thin white needles, by crystallizing
from petroleum ether, b. p. 80^0, at low temperatures.- The
melting point of this chloride was found to be 21.5^0-22.5^0un-
corr., that of the solid chloride 76^0-77^0 uncorr.-

Mr. S. R. McKee, working in this laboratory, has
recently shown that both anilides are formed by the action of
aniline on the pure low melting (or liquid) chloride in ether
solution.

Remsen and Kohler's work showed that the structure

(1) Am. Ch. Jour. 17, 330.
(2) Loc. Cit.
(3) Ibid, 17, 347.

of the fusible anilide is represented by the formula,

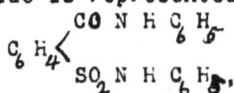

$$C_6H_4 \begin{cases} CO\ N\ H\ C_6H_5 \\ SO_2\ N\ H\ C_6H_5, \end{cases}$$

and that of the anil by the formula,

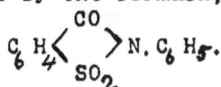

$$C_6H_4 \begin{cases} CO \\ SO_2 \end{cases} N.\ C_6H_5.$$

The evidence for these formulae is as follows:--

Orthotoluenesulphone chloride with aniline gives a compound of the structure

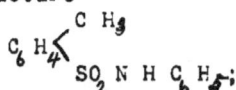

$$C_6H_4 \begin{cases} C\ H_3 \\ SO_2\ N\ H\ C_6H_5; \end{cases}$$

on oxidation this passes over to the acid,

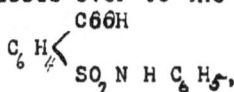

$$C_6H_4 \begin{cases} C6OH \\ SO_2\ N\ H\ C_6H_5, \end{cases}$$

which with phosphorus pentachloride goes over to the anil, which can hardly have any other structure than that indicated by the formula

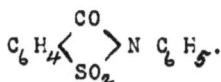

$$C_6H_4 \begin{cases} CO \\ SO_2 \end{cases} N\ C_6H_5.$$

Now this is identical in every particular with that obtained from the solid chloride, and when boiled with aniline it goes over into the anilide derived from that chloride, which there-fore has the formula
$$C_6H_4 \begin{cases} CO.\ N\ H\ C_6H_5 \\ SO_2.\ N\ H\ C_6H_5. \end{cases}$$

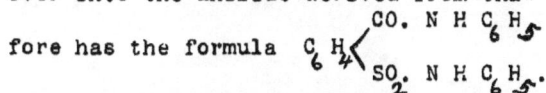

It follows therefore that, in all probability, this chloride

nas ~~been~~ the structure indicated by the formula

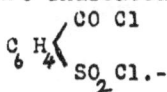

$$C_6H_4 \begin{cases} CO\ Cl \\ SO_2\ Cl. \end{cases}$$

The work of Remsen and Saunders (1) adduced additional evidence that this is the correct formula for this chloride.-

When concentrated ammonia is shaken with the mixed chlorides, the liquid chloride is readily decomposed, the solid chloride being but slightly attacked. Mr. A. D. Chambers of this laroratory, found that the product formed from the liquid chloride was ammonium orthocyan-benzenesulphonate,

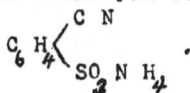

$$C_6H_4 \begin{cases} C\ N \\ SO_3\ N\ H_4 \end{cases} .$$

Mr. McKee has found that the pure liquid chloride forms the same compound, together with a small quantity of saccharine. It is difficult to understand the formation of this cyansulphonic acid under these conditions, except on the assumption of the formula

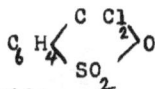

$$C_6H_4 \begin{cases} C\ Cl_2 \\ SO_2 \end{cases} O$$

for the liquid chloride.

If this be the correct formula for this chloride, the formula of the anilide derived from it follows, and should be written thus:--

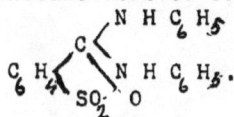

$$C_6H_4 \begin{cases} C \begin{cases} N\ H\ C_6H_5 \\ N\ H\ C_6H_5 \end{cases} \\ SO_2\ O \end{cases} .$$

(1) Loc.Cit.

Remsen and Kohler (1) found that this infusible anilide was a remarkably stable body towards reagents. The first
evidence obtained which indicated that there had not been a
deep-seated change in the arrangement of the constituent elements was from the reaction between it and benzoyl chloride,
by which benzanilide was formed.--The other products of this
reaction were not isolated.

They further ascertained that this anilide was acted
on by phosphorus oxychloride, and by phosphorus pentachloride,
though the products formed were not studied.

It was primarily with the hope of being able to clear
up the nature of these reactions that this investigation was
begun.-- The results arrived at are given below.-- For con-
venience the chlorides are spoken of as solid and liquid re-
spectively, and the anilides as fusible and infusible.

(1) Loc. Cit.

II. PREPARATION OF MATERIAL.

As the method now in use, devised by Mr. S. R. McKee, differs somewhat from that heretofore described for preparing the chlorides, it is given in detail.   The dry acid potassium salt prepared from commercial saccharine, is mixed, in finely powdered condition, with phosphorus pentachloride, in the ratio of one molecule of the salt to two and a half molecules of the pentachloride.   After the action has ceased, the phosphorus oxychloride is distilled off under diminished pressure, or in open dish on the water bath.-   The oily residue is then washed two or three times with cold water, by thoroughly agitating the mixture, in a wide-mouthed glass cylinder, with a conveniently bent glass rod.--   The oily chlorides, as free from water as possible, are then transferred to a beaker, placed in a freezing-mixture, and constantly stirred till the oil has solidified.   The stirring is necessary to prevent the solidifying mass from adhering to the beaker.-   In this solid condition, it is ground fine in a previously cooled mortar, transferred to a filter, and washed entirely free from the acids of

phosphorus by use of ice-water.-

If care be taken to avoid contact with any warm body, the chlorides may be pressed almost completely dry between folds of filter paper.-   They should be kept in a dessicator till required for use.-

III.   THE ACTION OF WATER ON THE MIXED CHLORIDES.

Remsen and Saunders (1) found that both chlorides were decomposed by water, forming the free acid together with hydrochloric acid, but   noted no difference in the rate of decomposition for the two substances.- To test this point, 2 grams of the mixed chlorides were placed in 150^{cc} cold water, and allowed to stand at the temperature of the room for 17 days, being thoroughly shaken from day to day.-   The insoluble residue was then filtered off, washed with cold water, and carefully dried.-   It weighed 0.41 gram, and, without crystallizing gave a melting point of 74°-75° uncorr.- Hence it was almost the pure solid chloride.-

IV. THE ACTION OF ANILINE ON THE CHLORIDES.

As already mentioned, Remsen and Coates, and later Remsen and Kohler also, studied the action of aniline on the mixed chlorides, and described the products formed.-- Thinking one of the chlorides might be more readily acted on than the other, the following experiments were made, with the results named below.-

1st.--To a weighed quantity of the chlorides in 100^{cc} of water was added aniline in the ratio of one molecule aniline to one of the chlorides.- After standing for 24 hours, with frequent shaking, crystals of each of the two anilides were obtained from the products formed, and some of the mixed chlorides had not been decomposed.- Aniline hydrochloride was also formed.

2nd.--In this experiment the ratio of the aniline to the chlorides was two molecules of the former to one of the latter, and the temperature of the water was kept at zero for the first three hours after mixing the compounds.- Besides aniline hydrochloride,both anilides were formed, and some of the solid chloride was still unacted on.- But from the experiment

with water and the chlorides, above described, the presence of
the solid chloride only may have been due to the fact that the
liquid chloride, not acted on by the aniline, had been decomposed by the water, as the products formed remained in the
water for 5 days before being examined.-

3rd.--The chlorides were next treated with aniline
in the ratio of four molecules of the latter to one of the
former, but ether was used as the menstruum, in place of water.
Much heat was given out during the reaction, the ether was kept
cool by means of cold water.- The products formed were the
two anilides and aniline hydrochloride.   Chloroform and benzene were tried as solvents in which to effect the reaction,
the relative quantities of the aniline and the chlorides being
the same as with ether as the solvent.-   In each case the
products formed were the same as when ether was used.-

As the yield seemed better when these solvents were
used than when the two compounds were brought together in emul-
sion in water, the use of the ether was adopted for preparing
the anilides used in this investigation.-   The method used is
given in detail:-

25 grams of the mixed chlorides are dissolved in absolute ether in a 500^{cc} Erlenmeyer flask; 38.9 grams aniline, also in solution in absolute ether, are added slowly to the chlorides, the solution being constantly stirred. The temperature of the solution is to be kept down by placing the flask in cold water.- A white precipitate appears as soon as any aniline is added. When the reaction is over, the ether is distilled off on the water bath, but this does not proceed smoothly, owing to the insoluble products of the reaction. To avoid bumping, the flask should be shaken constantly while the distillation proceeds. Even then bumping is not entirely avoided.-

The yellowish-white residue contains the mixture of the two anilides with aniline hydrochloride. Sometimes the residue has an oily appearance, and usually contains a slight excess of aniline. Cold water acidulated with hydrochloric acid is then added, the residue thoroughly stirred, and allowed to stand several hours. This treatment removes the aniline salts, and leaves the mixture of the two anilides.- Some of the infusible anilide will be dissolved in the water with the salts of aniline, hence this should be recovered by evapora-

ting the water to a small volume, when the anilide will crystalize, leaving the aniline hydrochloride in solution. This anilide is quite frequently colored, but may be purified by boiling with animal charcoal in dilute alkaline solution.-

To separate the two anilides the following method may be used:- The mixture is dissolved in hot alcohol, and this solution is cooled as rapidly as possible by putting the vessel in cold water. Under these conditions the fusible anilide crystallizes in thin white needles, while the infusible variety always crystalizes as short thick prisms. By rotating the vessel the fusible anilide will be disseminated through the alcohol, and may be poured off from the infusible prisms which will quickly settle to the bottom.-

By proceeding as above described there is usually formed a small amount of the anil, and sometimes a considerable quantity of it appears.- Since this is insoluble in cold dilute solution of sodium hydrochloride, while the anilides are soluble in that reagent, the latter may be readily separated from the former by that means.-

If the molecule of alcohol with which the infusible anilide crystalizes from solution in alcohol is undesirable,

the crystals should be dissolved in dilute solution of sodium
hydrochloride, and the solution acidified, when the anilide
will crystallize out without water or alcohol of crystallization.

V. ACTION OF PHOSPHORUS OXYCHLORIDE.

1. On the Infusible Anilide.- Remsen and Kohler (1)
observed that phosphorus oxychloride acted on the infusible
anilide, but none of the products were isolated.

Since phosphorus oxychloride attacks rubber or cork
stoppers, the following expedient was resorted to:-- A small
condenser was fitted over the neck of a tubulated retort.
Such an apparatus may be used for regular distillation, or
where a return condenser is desired, as was the case in this
experiment.   The infusible anilide is then boiled with an excess of the oxychloride, until all the anilide is dissolved.
This is attended by liberation of hydrochloric acid, and the
solution becomes quite yellow.   When the reaction is complete,
the excess of the   oxychloride is removed by distillation under diminished pressure.   There is left in the retort a light
brown viscous liquid.-   While still warm the retort is turned

(1) Am. Chem. Jour. XVII, 343.

about, so that this residue is spread over as much surface as

possible.- Water is then added, and the retort set aside un-

til the brown viscous mass has disappeared.   There remains

an insoluble almost white material, which dissolves in hot al-

cohol, forming a yellow solution.   After repeated crystalli-

zation, it melts sharply at 189.5° uncorr.   An analysis of

this product gave the following results:-

      I.   0.2002 gram gave 0.4997 gram CO_2 and

                 0.078 gram H_2O.

   II. 0.2015 gram gave 0.5020 gram CO_2 and 0.0795 gram H_2O.

   III. 0.2443 gram gave 0.1645 gram $Ba\ SO_4$.

   IV. 0.1683 gram gave 0.1132 gram $Ba\ SO_4$.

Calculated for				Found.				
$C_{17}H_{14}N_2SO_2$	I	II	III	IV	V	VI	VII
C	68.22	68.07	67.95	---	---	---	...	...
H	4.20	4.33	4.39	---	---	---	...	
S	9.59	---	---	9.24	9.25	---	---	---
N	8.41	---	---	---	---	8.63	8.61	8.5

      The combustions were made by the lead chromate meth-

od, as described by Te Roode(1), the sulphur determinations

(1) Am. Chem. Jour. XII, 226.

were made by the absolute method, exceptin, th. last one given
(VII), which Mr. E. McKay, in this laroratory, kindly made for
me by the method of Kjeldahl.-

Considerable difficulty was experineced in obtaining
satisfactory results for carbon, although the appearance and
the melting point of the compound indicated its purity. After-
wards a possible explanation was found in the fact that some
of the anil described by Remsen and Coates (1), was found to
result under the conditions of the experiemnt. Its formation
will be explained below.-

When pure this compound is of bright lemon-yellow
color, and the crystals obtained from alcohol are beautifully
formed arborescent plates, very brittle. It is only slightly
soluble in ether, $^{o^{nd}}$ chloroform, readily soluble in benzene, glacial acetic acid, and acetone.- When crystalized from either
of the last two solvents named, monoclinic prisms of great
brilliancy are obtained, the faces developed being usually ∞P,
$\infty P\bar{\infty}$, o P, P, and $2P\infty$.-

2. On the Fusible Anilide.- This same compound is
formed by the action of phosphorus oxychloride on the fusible
anilide, the details of the experiment being the same as with

the infusible anilide, given above.   Some anil is found in
this case also.-

Efforts to get a crystalized product from the brown
viscous residue, by using various solvents, were unsuccessful.
It appears that this residue is really a solution of the yellow product in the phosphoric acid, from the following consid-
eration:--   To a small amount of phosphoric acid prepared by
adding very little water to phosphoric anhydride, was added
some of the pure yellow product, and the mixture was then
heated.-   A clear solution was effected, having the same con-
sistency and color as the residue obtained above;   and when
water was added to this, it decomposed in much the same way.-

If the empirical formula arrived at by analysis of
this yellow product be compared with that of the anilide from
which it is derived, the equation,

$$C_{19} H_{16} SO_3 N_2 - H_2 O = C_{19} H_{14} SO_2 N_2 ,$$

shows their relation.   It thus appears as if the action of
the oxychloride is one of dehydration.

VI. THE ACTION OF PHOPHORUS PENTOXIDE.

1.On the Infusible Anilide.-   It seemed reasonable to

expect the same compound to result if either of the anilides should be treated with a strong dehydrating agent.   Accordingly, some infusible anilide was mixed with phosphorus pentoxide in a test-tube and then heated for two hours in an air-bath to 130°-150°, the test-tube being closed by a Bunsen valve. The mass had become slightly yellow in color. When cold, water was added to dissolve the phosphoric acid and anhydride, and the portion insoluble in water was found to have the same crystalline form, when deposited from alcohol, and to melt at the same temperature, as the product from the other experiments already described, using the oxychloride of phosphorus.

2. On the Fusible Anilide.- If the fusible anilide be treated with phosphorus pentoxide, as just described, the same yellow compound is formed.-

Neither acetyl chloride nor acetic anhydride effect this dehydration of the anilide.-

Hoogewerff and Van Dorp (1) have described the formation of some iso-imides, by treating dibasic acid derivatives of the general formula,

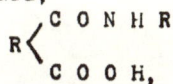

$$R \bigg\langle \begin{array}{l} CONHR \\ COOH, \end{array}$$

with either acetyl chloride, or phosphorus oxychloride, the
reaction with the former being represented by the equations,

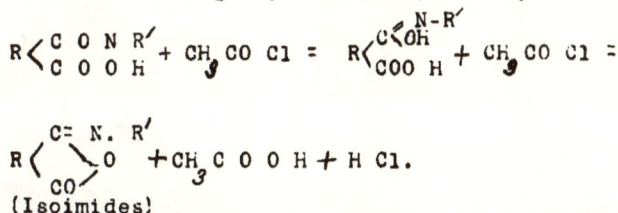

$$R \Big\langle {}^{C\ O\ N\ R'}_{C\ O\ O\ H} + CH_3\ CO\ Cl\ =\ R \Big\langle {}^{C \langle {}^{N-R'}_{OH}}_{COO\ H} + CH_3\ CO\ Cl\ =$$

$$R \Big\langle {}^{C\dot=\ N.\ R'}_{\ \ \ O} \Big/_{CO} + CH_3\ C\ O\ O\ H + H\ Cl.$$
(Isoimides)

With the oxychloride the reaction would be analo-
gous.-   In all cases examined, $\underline{R'}$ was CH_3 , C_2H_5, or $C_6H_5CH_2$,
while $\underline{R}$ was the residue of camphoric, and phthalic acids re-
spectively.   Clearly the role of the oxychloride in these reactions was the same as that which it plays in the reaction
with the anilides.   The equation,

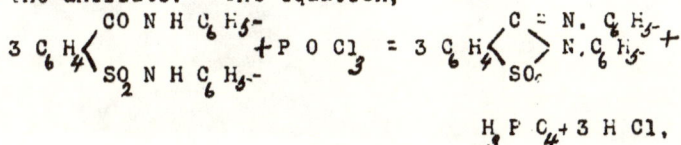

$$3\ C_6H_4 \Big\langle {}^{CO\ N\ H\ C_6H_5-}_{SO_2\ N\ H\ C_6H_5-} + P\ O\ Cl_3\ =\ 3\ C_6H_4 \Big\langle {}^{C\ =\ N.\ C_6H_5}_{SO_2} \Big\rangle N.C_6H_5 +$$

$$H_3\ P\ C_4 + 3\ H\ Cl,$$

then represents the reaction with the fusible anilide.-

3. Structure of this Compound.- From the methods of
formation already described, especially those involving the
use of the fusible anilide, the formula for which is entirely
satisfactorily represented as above, it naturally follows
that the formula,

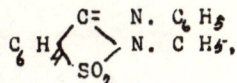

$$C_6H_4 \Big\langle {}^{C\dot=\ N.\ C_6H_5}_{SO_2} \Big\rangle N.\ C\ H_5,$$

correctly represents the structure of the compound. It thus
appears to be a di-anil of orthosulphobenzoic acid.

Jesurun (1) obtained this same compound by the action of aniline on orthocyanbenzenesulphone chloride. He thinks the first product formed is a benzamidine-sulphanilide,

$$C_6H_4 \begin{cases} C \diagup\!\!\diagup N.C_6H_5 \\ \diagdown N H_2 \\ SO_2 N HC_6H_5 \end{cases}$$

and that this loses ammonia, forming this dianil, to which he gives the same structural formula as above assigned.-

The grouping of the two aniline residues in this dianil is the same as that given by Auschütz and Beavis (2) for the dichlormaleic-dianil, the formula of which is,

$$\begin{matrix} C.Cl- C \diagup\!\!\diagup N.C_6H_5 \\ \| \qquad \diagdown N.C_6H_5, \\ C.Cl- CO \end{matrix}$$

which is formed by the action of aniline on dichlormaleic-anil-dichloride,

$$\begin{matrix} C Cl. C Cl_2 \\ \| \qquad \diagdown N.C_6H_5, \\ C. Cl-C:O \end{matrix}$$

prepared by the action of phosphorus pentachloride on succir anil (3).

It seems reasonable to expect the formation of the analogous derivatives of the anil described by Remsen and Coates. Preliminary experiments have been made with this end in view, but as yet without success. The matter is still under investigation.-

4. Chemical Properties of the Di-anil.- (a) It was mentioned above that this compound may be crystalized from glacial acetic acid.- For this purpose the solution must not be heated longer than is necessary to effect solution. If the solution be boiled for two or three hours in a flask connected with a reflux condenser, none of the dianil will separate on cooling, but instead there will appear slender white needles.- These are crystals of the infusible anilide. The identity was established by the melting point, crystalline form when recrystallized from alcohol, and from water, ready solubility in dilute cold alkaline solution, from which the characteristic granular crystals of the infusible anilide separate when the solution is acidified. The infusible anilide itself assumes the acicular habit if crystallized from glacial acetic acid.- To further test the identity, some of these needles were treated with phosphorus oxychloride, and the dianil, with

some of the ordinary anil, was obtained, just as when the infusible anilide was used.-

This conversion of the dianil into the infusible anilide is accompanied by a change in color of the acetic acid solution, from a darker to a much lighter shade of yellow.- This reaction is represented by the equation

$$C_6H_4\underset{SO_2}{\overset{C\,=\,N.\,C_6H_5}{\diagup\!\!\!\diagdown}}N.\,C_6H_5 + H_2O = C_6H_4\underset{SO_2}{\overset{C\overset{N\,H\,C_6H_5}{\underset{N\,h\,C_6H_5}{\diagdown}}}{\diagup\!\!\!\diagdown}}O.$$

The yield is quantitative.

(b) The crystals of the dianil may be boiled with a water solution of sodium hydroxide, either dilute or strong, without being affected in any way.- When boiled for about an hour with alcoholic potash, a solution of the dianil is effected. If now the alcohol be evaporated almost completely, there is no precipitate formed on addition of water. If, however, this solution be acidified, crystals of the infusible anilide separate on standing, provided the solution be sufficiently dilute. If the solution be quite concentrated, the anilide will be precipitated at once by the acid.

The effect of boiling the dianil with alchoholic potash is to induce the dianil to unite with the elements of water, to form the infusible anilide.- It is worthy of note that

these elements arrange themselves in the more stable grouping

of the infusible, rather than the fusible anilide, which is

less stable.

      (c) When concentrated hydrochloric acid is added to

the dianil, the xxkkxx color of the latter changes from yellow

to white.   If the acid be filtered off at once, and the resi-

due washed with cold water, the yellow color reappears. If,

however, the dianil be boiled with the acid for about an hour,

the insoluble residue will be found to remain white after re-

moving the acid completely, and when crystallized from alcohol,

it forms slender white needles.   These are crystals of the

anil described by Remsen and Coates (1).   In the hydrochloric

acid solution is some aniline hydrochloride.   This reaction

may be represented by the equation,

$$C_6H_4 \begin{matrix} C = N.C_6H_5 \\ N.C_6H_5 \end{matrix} + H_2O + HCl = C_6H_4 \begin{matrix} CO \\ SO_2 \end{matrix} N . C_6H_5 + C_6H_5NH_2 . HCl.$$

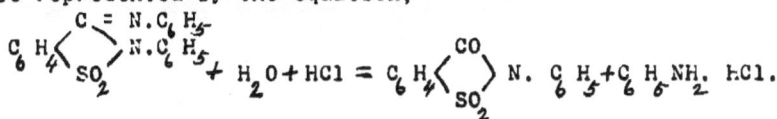

It was mentioned above that some of this anil was always formed

during the action of phosphorus oxychloride on the anilides.

In the light of the above reaction, its presence is explained,-

being a secondary reaction between the dianil and the hydro-

chloric acid liberated during the reaction.

(1) Loc. Cit.

VII.-ACTION OF PHOSPHORUS PENTACHLORIDE ON THE INFUSIBLE ANILIDE.-

When the infusible anilide is added slowly to a
heated solution of phosphorus pentachloride in chloroform, re
action sets in at once, and generally the solution changes to
a light yellow in color. Usually about twice as much pentachloride as of the anilide is necessary to effect complete solution of the latter.-   In the first experiment tried, 2.5
grams of the pentachloride were used to one gram of the anilide,
and the heating was continued only as long as was necessary to
effect complete solution of the anilide.- On cooling, radia-
ting tufts of fine white needles separated, completely filling
the liquid.-   More chloroform was added, and heat applied till
all the needles were dissolved.- But the needles did not separate again, even when almost all the chloroform had evaporated.
The last traces of chloroform were then driven off by heating
gently for some time, and the yellow viscous residue was treated as the brown residue from the oxychloride experiment (p.-12)
Some of the anil was found present, though the chief product
was the yellow dianil.-   The final products of this reaction

are the same as those formed by the action of phosphorus oxychloride on the infusible anilide.-

This experiment was repeated under the same
condition as before, and under varied conditions, though always
without obtaining those fine white needles. The final products were the same each time.

From the results of the experiments it will readily
be seen that it is possible to pass from the fusible anilide
to the infusible anilide, by passing first to the yellow dianil, and then, boiling this with glacial acetic acid, or alcoholic solution of potassium hydroxide, to the infusible
anilide.

Again, from this dianil one may obtain the anil
by boiling with concentrated hydrochloric acid. And since
the anil, when boiled with aniline, passes over to the fusible
anilide, it is therefore possible to pass in the opposite di-
rection,-from the infusible to the fusible.

VIII.-THE ACTION OF BENZOYL CHLORIDE ON THE INFUSIBLE ANILIDE.-

Five grams infusible anilide were boiled with an excess of benzoyl chloride until it had dissolved. The excess

of the benzoyl chloride was removed as completely as possible
by distillation under diminished pressure, and the residue
boiled with water.   After this water had become cold, and
without separating it from the insoluble residue, sodium carbonate was added to alkaline reaction, and the vessel was heated gently on the water-bath to about 60° C.-   This solution
was then poured off from the undissolved residue, and acidified,
when benzoic acid was precipitated.-

The residue insoluble in sodium carbonate solution
was then dissolved in alcohol, and boiled with animal charcoal.
From this solution separated plates of a dirty-yellow color,
but which when purified by repeated recrystallization became
white, and assumed the acicular habit, melting sharply at
190.5° uncorr.   They proved to be the anil.-

The mother liquor from the above was then evaporated
to dryness, and the residue boiled with about 750 c.c. water,
and this water while still hot was siphoned from the part insoluble in the water, which had melted to a brown oil.   By
repeatedly boiling this residue with water, almost all of it
was finally dissolved.

From this water, on cooling, a white precipitate was

deposited.- When crystallized from alcohol, this forms small
globular shaped tufts of white needles, but these have not yet
been identified, partly from lack of material, partly from lack
of time.

From the water filtrate from this white precipitate,
after eveporating to a very small volume, a small quantity of
benzanilide was obtained, which was identified by its melting
point, crystal form, and by comparison with a specimen of benzanilide.- The equation,

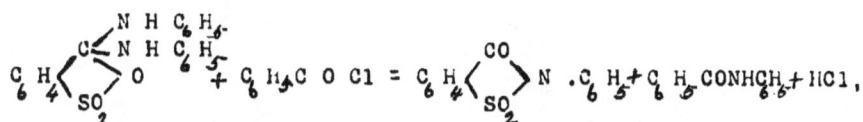

$$C_6H_4 \underset{SO_2}{\overset{C\underset{N\,H\,C_6H_5}{\overset{N\,H\,C_6H_5-}{<}}}{<}} O + C_6H_5C\,O\,Cl = C_6H_4 \underset{SO_2}{\overset{CO}{<}} N\cdot C_6H_5 + C_6H_5CONHCH_6 + HCl,$$

represents the reaction as far as yet understood.-

A second experiment gave the same results as the one
just described.- No trace of the product mentioned by Remsen
and Kohler (1), as giving rise to the infusible anilide on addition of aniline, was obtained in either case.- It is possible that this product might be isolated by keeping the temperature as low as possible in order to effect the reaction.

STRUCTURE OF THE INFUSIBLE ANILIDE.

From the results obtained in this investigation, it

(1) Am. Chem. Jour. 17, 342.

is clear that the two aniline residues are intact in the infusible anilide.   This fact, together with the further fact that
this anil$\overset{\cdot\text{ide}}{\,}$ is derived only from the liquid, or unsymmetrical
chloride, by the action of aniline, leaves but little if any.
doubt that the structure of this anilide is correctly represented by the formula,

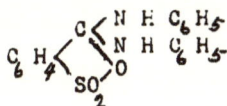

$$C_6H_4\Big\langle\begin{matrix}C\big\langle\begin{matrix}N\ H\ C_6H_5\\N\ H\ C_6H_5\end{matrix}\\SO_2\end{matrix}O$$

first proposed by Remsen and Coates.-(1)

IX.- CONCLUSION.

The results arrives at in this investigation may be
briefly summarized as follows:-

1. Water decomposes the liquid chloride more readily
than the solid chloride.

2. Orthosulphobenzodianil, having the structure represented by the formula,

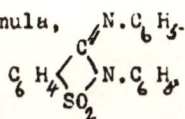

$$C_6H_4\Big\langle\begin{matrix}C\big\langle\begin{matrix}N.C_6H_5\\N.C_6H_5\end{matrix}\\SO_2\end{matrix}$$

is obtained by the action of phosphorus oxychloride, or phosphoric anhydride on either anilide, and by the action of phos-

(1)Am. Chem. Jour. 17, 320.

.horus pentachloride on the infusible anilide.

3. The infusible anilide is re-formed from this di-anil on boiling with glacial acetic acid, or with alcoholic potash.

4. The anil of the orthosulphobenzoic acid is obtained by boiling this dianil with concentrated hydrochloric acid.

5. Benzoyl chloride acts on the infusible anilide, forming benzanilide and the anil of orthosulphobenzoic acid.-

6. The two aniline residues are intact in the infusible anilide, and this anilide has the structure represented by the formula,

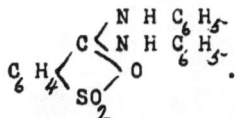

$$C_6 H_4 \left\langle \begin{array}{c} C \left\langle \begin{array}{c} N\ H\ C_6 H_5 \\ N\ H\ C_6 H_5 \end{array} \right. \\ SO_2 \end{array} \right\rangle O \; .$$

BIOGRAPHICAL.

The author was born near Apex, Wake Co., N. C. June 6th, 1865. His early education was received in private schools in that village. He entered Wake Forrest College, N. C. September, 1881, receiving the degree of Bachelor of Arts in 1885. For three succeeding years he taught in private schools in his native State. He returned to Wake Forrest College September, 1888, and received the Master's degree in 1889. In October of that year he entered the Johns Hopkins University, remaining two years. During the two years 1891-93, he taught in the Wisconsin State Normal School, located at Oshkosh.- Since October, 1893, he has continued his work at the Hopkins University. His subjects have been Chemistry, Mineralogy and Physics.